W9-ASL-822

ROLYPOLYOLOGY

ROLYPOLYOLOGY

by Michael Elsohn Ross
photographs by Brian Grogan
illustrations by Darren Erickson

Carolrhoda Books, Inc. / Minneapolis

To my wife, Lisa, who inspired me with stories about the rolypoly family she once kept in a shoebox under her bed.

Text copyright © 1995 by Michael Elsohn Ross
Photographs copyright © 1995 by Brian Grogan
Illustrations copyright © 1995 by Carolrhoda Books, Inc.

Carolrhoda Books, Inc. c/o The Lerner Publishing Group
241 First Avenue North, Minneapolis, MN 55401 U.S.A.

Website address: www.lernerbooks.com

LIBRARY OF CONGRESS CATALOGING-IN-PUBLICATION DATA

Ross, Michael Elsohn, 1952–
    Rolypolyology / Michael Elsohn Ross ; photographs by Brian Grogan ;
    illustrations by Darren Erickson.
        p. cm. — (Backyard buddies)
    Includes index.
    ISBN 0-87614-862-3
    1. Isopoda—Juvenile literature. 2. Isopoda as pets—Juvenile
literature. 3. Zoology—Study and teaching—Activity programs—
Juvenile literature. [1. Isopods.] I. Grogan, Brian, 1951–
ill. II. Erickson, Darren, ill. III. Title. IV. Series: Ross,
Michael Elsohn, 1952–   Backyard buddies.
QL444.M34R67 1995
595.3'72—dc20   94-22327

Manufactured in the United States of America
2 3 4 5 6 7 – JR – 03 02 01 00 99 98

Contents

If I could curl up in a ball,

I'd roll down the hall.

My house would be small.

I'd live in myself.

I'd be a rolypoly.

Miniature Armadillos

Have you ever seen one of these before?

It is called a rolypoly because it can roll up into a ball.

Some people think rolypolies look like little pills, so they call them pill bugs. Other folks call them sow bugs, bibble bugs, tiggy hogs, or potato bugs. Rolypolies are also known as wood lice. But they aren't bugs, lice, or hogs. Though they are similar to insects, they belong to a special group of animals known as **isopods.** There are many kinds of isopods, and only some roll up in a ball like an armadillo. In fact, Armadillididae is the scientific name of one family of rolypolies.

Geology is the study of the earth. Anthropology is the study of people. Rolypolyology is the study of rolypolies. Rolypolyologists are explorers who learn about their rolypoly buddies by watching them. They are also scientists who study living things with care and concern about the lives of small creatures. To practice rolypolyology, you do not need to hurt or kill anything. All you need is plenty of curiosity and a touch of patience. Keep reading and you can become a rolypolyologist.

Rolypoly Hunt

Are you ready to find some rolypolies? First, get a cup or other small container. Then try looking under rocks, boards, piles of dead leaves, or logs. Most rolypolies like to live in moist places, but there are some that live in deserts. Rolypolies even live in cities. A community garden or vacant lot might be a good place to search for them.

Don't worry about rolypolies hurting you. They can't bite or sting. But you can hurt them if you're not careful. Just because you're 100 times bigger doesn't mean that you can't be considerate. Be a gentle giant. A leaf or small stick can be used to steer or roll rolypolies into the container. Be sure to keep your rolypolies in a shady place. Long exposure to direct sunlight will dry them out and eventually kill them.

Before you bring the rolypolies into the house, think about your parents. How will they react? If they are the kind of people who sometimes get squeamish about little critters, perhaps you can soothe them by showing them the following article from the *Daily Planet* (a well-respected fictional newspaper).

The Daily Planet

The Ultimate Pet

Pearl Perez, of Lindenhurst, finally found the solution to the pet problem. After going through two dogs, a cat, and tropical fish, her family has finally settled on the gentle creatures that her two children, Patty and Paco, discovered in the backyard. Mrs. Perez admits that she was at first a little frightened by the tiny rolypolies, but Patty and Paco assured her that rolypolies are harmless, gentle beings. Mrs. Perez was still wary, but she showed courage by telling the kids they could keep them for a short time. Within a few days, Mrs. Perez realized that not only were her children right about the rolypolies being harmless, but also that they had finally solved the family's "pet problem." Unlike dogs and cats, rolypolies do not require expensive food, veterinarian visits, or housebreaking. They don't bark, tear up the upholstery, or chew on shoes. They never drool. Tropical fish cost a fortune, but rolypolies are free for the taking.

Mrs. Perez is grateful to her children for finding such wonderful pets and has this to say: "It was the best thing that ever happened in our home. I mean, I'm happy, my kids are happy, and I think the rolypolies are happy too."

It's fun to have rolypolies as houseguests. They aren't noisy, they don't complain, and they are fascinating to watch. Like all visitors, they need safe, comfortable lodging. If you are ready to host a rolypoly convention, you can make a simple hotel with the following materials:

container, you won't have to worry about the rolypolies escaping.

3. Your hotel is now ready for guests, so invite some rolypolies for a visit.

4. Make a sign and tape it to the hotel if you want to inform other people, such as your parents, that you have houseguests.

Warning: Direct light and heat can be hazardous to the health of rolypolies. Locate the hotel in a cool, shady neighborhood.

The leaves and block will provide a good place for the rolypolies to hide. If the hotel becomes too dry, they may burrow down into the soil. An occasional sprinkle of water will keep the soil moist; any more water may cause a flood.

Hotels usually aren't permanent residences. After your rolypolies have been visiting for a while, be a considerate host by returning them to their real backyard home.

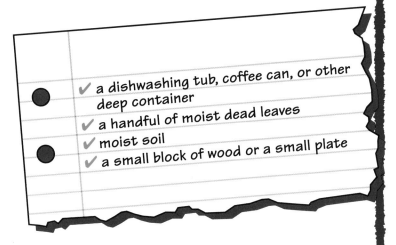

✔ a dishwashing tub, coffee can, or other deep container
✔ a handful of moist dead leaves
✔ moist soil
✔ a small block of wood or a small plate

How to Make the Hotel:

1. Fill the tub with soil to within 2 to 3 inches of the rim.

2. Set the block or plate on top of the soil and scatter the leaves around the block. If the soil and leaves are several inches below the lip of the

How are your rolypolies behaving? Even though it's not polite to stare, it's okay to watch your guests. Pretend you are a house detective. Get a notepad and pen. Settle down where you can observe in comfort and watch your subjects' every move. If you can't see any, check for them under the block of wood.

What are they doing? Jot down some notes about their behavior. You never know when these observations might come in handy. Are there any patterns to their behavior? Are they doing anything? Are you falling asleep?

Detectives sometimes have to sit in their spy posts for hours watching their suspects. How long can you spy? Of course, it's sometimes easier with a little help from your friends. Perhaps you can find a friend to assist you with the surveillance.

Be a Hotel Detective

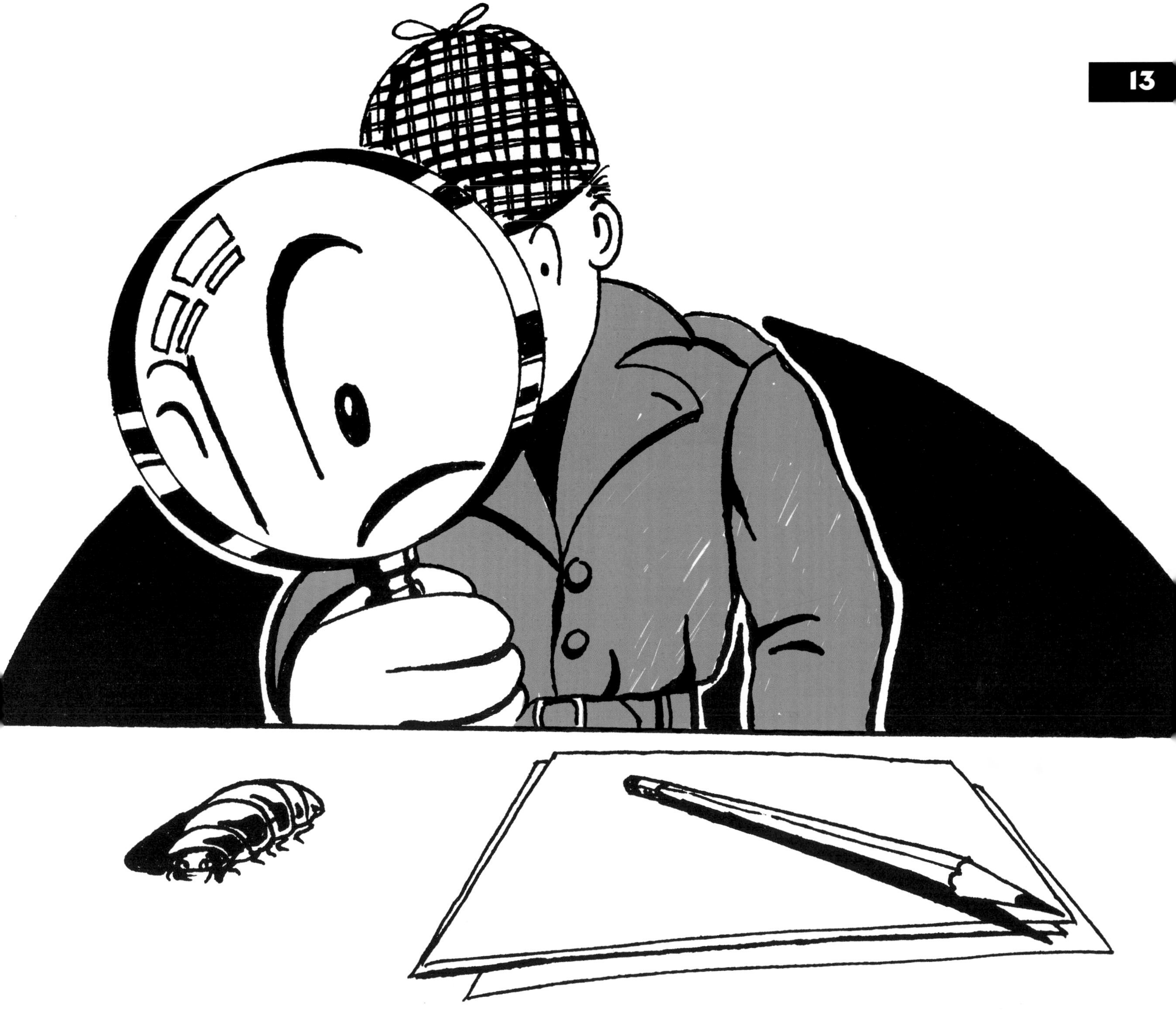

Some hamburger joints have playgrounds. Why not make a playground for your hotel? You can construct a great miniature playground in a large baking pan with some of the following items. (Be sure to get your folks' permission before using their stuff!)

Rolypoly Playground

Place a rolypoly in the playground. What does it do? Take out your detective notebook and jot down some observations. Gently touch the rolypoly. How does it react? Does it curl up in a ball? Try placing a few more rolypolies in the playground. What do you think will happen?

Have you ever thought of yourself as a giant jungle gym? To become one, all you need to do is place a rolypoly on your hand. Sit still and watch. Can it climb up your arm? Over your arm hairs? Does it tickle?

When playtime is over, be sure to return the rolypolies to their hotel. They'll be safer there than wandering around the kitchen.

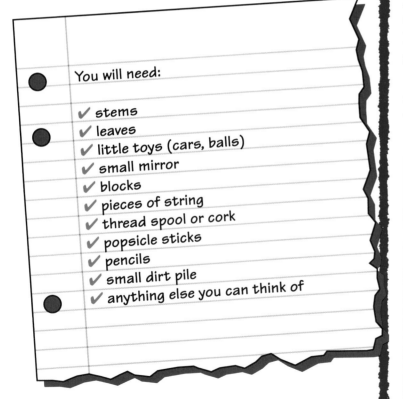

You will need:

- ✔ stems
- ✔ leaves
- ✔ little toys (cars, balls)
- ✔ small mirror
- ✔ blocks
- ✔ pieces of string
- ✔ thread spool or cork
- ✔ popsicle sticks
- ✔ pencils
- ✔ small dirt pile
- ✔ anything else you can think of

You can play this game alone, but it is easier with two or more partners.

Seeing Game

ball," "have legs", etc.).

4. As each player finishes describing a feature of the rolypoly, the next player takes a turn. Continue taking turns in order until a player is unable to make a new observation. Any observation is okay, but no repeats are allowed. However, you can add more details to a previous observation. For example, someone may say, "It rolls into a ball" and another person can say, "It rolls into a ball when it is touched."

See how many observations you can make. The last person able to make a new observation is the Champion Seer.

You will need:

✔ a bowl
✔ a magnifying lens
✔ optional: pen or pencil and paper

How to Play:

1. Place a rolypoly in the bowl and sit near it.

2. When playing with several people, decide who goes first.

Optional: pick one player to write down observations.

3. The one who starts the game says, "I notice that rolypolies _____" (fill in the blank with your observations—for example, "roll in a

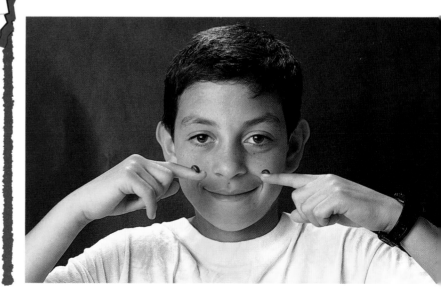

Living things are mathematical. People, for example, have body parts in sets. We have two hands, ten fingers, and twenty nails. Our body parts also come in a variety of shapes. Our eyes are round, and our noses are triangular. Would you like to check out the mathematics of a rolypoly?

Rolypoly Math

Here's what you need for a rolypoly examination:

✔ a magnifying lens
✔ a piece of masking tape
✔ a bit of straw or a blade of grass

Make a one-inch-long loop with the masking tape. Place the tape on a table and stick the back side of a rolypoly to the tape. (If the rolypoly is curled up, just wait a few minutes until it uncurls

itself.) This won't harm the rolypoly as long as the tape only touches its shell. While the rolypoly is stuck, check it out with the magnifying lens. Does it have legs? How many? How about eyes and feet? Can you find its mouthparts? Can you see how it is able to curl its body? Count any other parts you can find and see if there are any patterns to the numbers.

How many different colors are you wearing today? How many colors is the rolypoly wearing? Is there more than one? Besides colors, can you count shapes? How many different shapes do you see? What are they?

This activity will help you notice even more rolypoly details. In fact, you may discover something that no one else has ever seen.

Portrait of a Giant

and ant eyes, you can show people that there is a whole lot more to them. As you peer through the lens, pretend that your eyes are tiny little ants crawling over a giant creature. Each time your ant eyes walk over a new part of the rolypoly, use the pencil to jot down a quick sketch of what you see. Don't be concerned with putting all the parts together yet, and don't worry if your pictures aren't exactly perfect. The more you sketch, the easier it will become.

Do you wonder about some of the things that you see? If you have a question, jot it down. Questions are a sign that you are thinking, and thinking about rolypolies is fun.

Once you finish drawing rolypoly parts, you may want to sketch some rolypoly outlines. What does it look like from above, from below, or from a side view?

You are now ready to create a magnificent rolypoly poster. Lightly sketch a page-size outline of a rolypoly. Add details from your other sketches. Perhaps you can even toss in some color. When you are finished working on the poster, hang it on your wall. Later on, if you notice more rolypoly details, you can just add them to your poster. If you continue to study rolypolies, who knows—you may end up working on your isopod portrait for another thirty years.

What You Need:

✔ a pencil and several sheets of scrap paper

✔ a magnifying lens

What to Do:

Imagine a rolypoly four stories high. What would people do when they saw it coming? Would they call in the air force? Take a look at a rolypoly through a magnifying lens, and it may appear gigantic to you.

When most people look at a rolypoly, all they see is a little gray bug, but with a magnifying lens

Wondertime

Do rolypolies have eyes? Can they swim?

Are they in the tick family? Do they have teeth?
Do they have ears? Do they have babies?

Do younger ones have fewer legs? How far can they go in 30 seconds?

What kind of food do they like? How long do they live?

How steep can they climb? Do they like light or dark?

Why can't the flat ones roll up? Are the flat ones girls?

Are the rolled-up ones boys? How many bones do they have?

Do they come in other colors? Are they insects?

Why are the smaller ones brown and the bigger ones gray?

Do rolypolies roll to get from one place to another?

Do they have jaws? Where are their gills?

What are the white things underneath?

These are some questions that kids in my town asked about rolypolies. Do you have any answers for them? Do you have any questions of your own? If you do, don't let them get away! Many great adventures have started with a simple question. For example, think of the people who asked "Is the world round?" or "What's it like on the Moon?" Record your questions and get ready for some wild journeys.

Are you ready for a potentially wild and crazy challenge? If you answered yes, all you have to do to get started is to focus on a rolypoly question. Is there something you really wonder about rolypolies? Yes? Well, take that question and follow it. The tips below may help you stay hot on the trail.

Follow That Question

— **Scrutinize:** Do you think you could answer your question if you did a little more observation? For example, if your question was "Do they have eyes?" do you think you might be able to find the answer if you examined a rolypoly closely through a magnifying lens?

— **Find an Expert:** Do you know any people who know a lot about bugs and other little creatures? Do you think they might be able to answer your rolypoly question? Consult a neighborhood gardener, plant expert, or local science teacher. For more difficult questions, try a **zoologist** (a scientist who studies animals) at the local college, science museum, or zoo.

— **Research:** Other rolypolyologists may have pondered your question in the past and discovered an answer. Perhaps you could check out some books. In fact, a great book to refer to is

this one. Turn the page and search through the next section. If that doesn't answer your question, come back to this page and read on.

— **Experiment:** Experimentation is one road to discovery. Check out the chapter on Kid Experiments to get some ideas about how other people pursued their questions about rolypolies. Could you design an experiment to explore your question?

What defends itself like a skunk and an armadillo? What carries its babies like a kangaroo? What eats like a bunny? Would you believe a rolypoly?

Rolypolies are similar to all of these animals in many ways, but their closest relatives are lobsters, crabs, and shrimp. Do you

What Is It?

wonder how a rolypoly can be related to expensive seafood? Well, for one thing, all have crusty outer skeletons. Perhaps that's why they are all called **crustaceans** (krus-TAY-shunz). Insects are also crusty, but they have three separate body parts. Most crustaceans have only two body parts, the **cephalothorax** (seh-fuh-luh-THOR-aks)

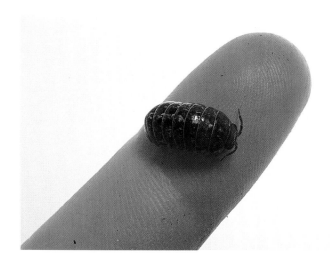

and the **abdomen** (AB-duh-men). *Cephalo* is Greek for head, and the cephalothorax is where the head is located. The heads of many crustaceans are protected by a hardened covering called a **carapace** (KAR-uh-pus).

Tap the shell-like plates that cover a rolypoly's back. Are they hard? Look at the head. Can you see the hood-like carapace that covers it? Both the carapace and plates

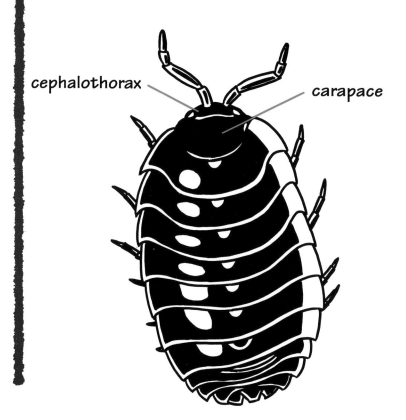

cephalothorax

carapace

are part of a rolypoly's external skeleton. There are no bones inside. Like other creatures who wear hard outside shells, rolypolies can get a little crowded inside their **exoskeletons** (ek-so-SKEH-luh-tunz). When this happens, they simply shed their old ones and replace them with bigger ones.

All crustaceans, including rolypolies, have many pairs of jointed legs and **segmented** bodies. They also wear two pairs of **antennae** (an-TEH-nee), which they use for sensing things around them. The first set, called **antennules** (an-TEN-yoolz), are so small that they are hard to see without magnification. Right behind the second set of antennae are a pair of mouthparts called **mandibles** (MAN-duh-bulz).

Rolypolies belong to a special group of crustaceans, the isopods. *Isopod* means equal foot. Unlike crabs, lobsters, and shrimps, an isopod's legs are all the same shape. Most isopods live in water, but rolypolies are landlubbers.

Since rolypolies are crustaceans, they are similar to other bug-like creatures such as centipedes, millipedes, insects, and spiders. All of these animals belong to a larger group known as the **arthropods**. *Arthro* means joint, and *pod* means foot. All arthropods have jointed feet.

Look at your rolypoly and your rolypoly sketches. Does your rolypoly fit the description of an isopod, a crustacean, and an arthropod?

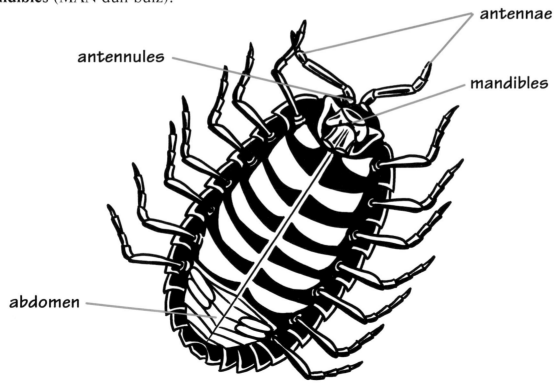

antennae

antennules

mandibles

abdomen

Each language has its own name for rolypolies. In Spanish, the word for rolypoly is *chicharra*. In Hindi, it's *koojehlee kira*. In Dutch, it's *pissebed*. Is it a chicharra, koojehlee kira, pissebed, or rolypoly? Since having lots of names for one creature can get a little confusing, scientists have invented an international system of naming all living things. Whether you live in Mexico, India, or Indonesia, there is only one name for each species of rolypoly.

Latin and Greek are used in creating scientific names. The name *stegosaurus*, for example, is composed of two Latin words, *stego* (which means "covered") and *saurus* (which means "lizard"). In fact, most of us only know dinosaurs by their scientific names! On page 26, you will find scientific names for rolypolies.

Chicharra... Say What?

Rolypolies are part of a group of animals called isopods.

Isopods are part of a group of animals called crustaceans.

Crustaceans are part of a group of animals called arthropods.

Arthropods are all creatures with paired jointed legs. The animals below are all arthropods.

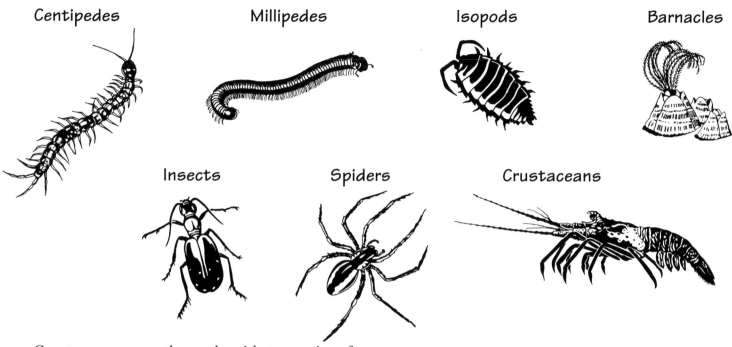

Centipedes

Millipedes

Isopods

Barnacles

Insects

Spiders

Crustaceans

Crustaceans are arthropods with two pairs of antennae. The animals below are all crustaceans.

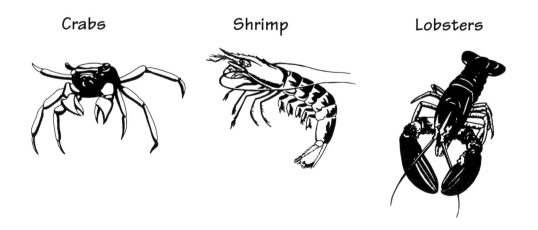

Crabs

Shrimp

Lobsters

Rolypolies are isopods, but not all isopods are rolypolies. An isopod that rolls up into a ball is called a rolypoly or pill bug. A land isopod that does not roll up is called a sow bug. If it lives along the shore, it's a rock slater. Water slaters are isopods that live in the water.

Isopods

The largest water slater is a lilac-colored resident of deep waters that may grow longer than this page! There are over four thousand different species of isopods in the world. Most are smaller than a quarter.

	Shore	**Forest**	**Grass**	**Semi-dry**	**Desert**

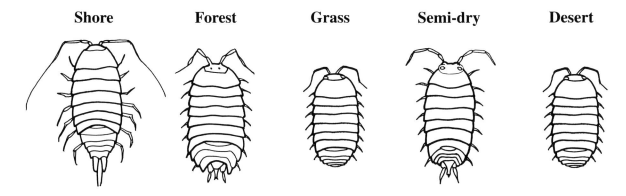

Family Name	Ligiidae	Oniscidae	Armadillididae	Porcellionidae	Armadillidae
Can they roll up?	No	No	Yes, but not antennae	No	Yes, including antennae
Need for moist home	High	High	Medium	Medium	Low

As you can see from the chart, only two families of isopods roll up in a ball. Members of the family Armadillididae (ar-muh-dih-LID-uh-dee) can roll up everything except their antennae, while rolypolies in the family Armadillidae (ar-muh-DIH-luh-dee) can roll up everything. When the common rolypoly *(Armadillidium vulgare)* curls up, you can still see its antennae. What family do your rolypolies belong to?

Common Rolypoly
Family: Armadillididae
Genus: Armadillidium
Species: Vulgare (meaning "common")

The backyard garden can be a wild and dangerous place for small critters such as rolypolies. Birds, lizards, toads, salamanders, spiders, and other small garden **predators** are constantly on the prowl for tasty snacks. Fortunately, rolypolies have numerous ways of protecting themselves. The most obvious is their habit of curling up in a ball like an armadillo. Making a stinky odor like a skunk is another method used to foil enemies. Kids are supposed to have a good sense of smell. Take a whiff of a rolypoly. Does it stink?

Often rolypolies curl up into a ball when they are climbing plant stems or rocks, and the result is that they quickly roll out of sight. Could rolling away be another way to escape predators? In Europe, a species of rolypoly has been discovered that looks like the European black widow spider. When the rolypoly is curled up, it has the same colors and markings found on the abdomen of the black widow. Black widows taste nasty. Animals that survive a black widow meal most likely never want to try one again. Looking like a black widow could be a real lifesaver.

In the end, perhaps a rolypoly's best defense is its fondness for dark, sheltered crannies and other hard-to-reach hiding spots. In these remote places, rolypolies are safe from bigger predators, such as lizards, toads, and birds. In very narrow crevices, they are even protected from deadly spiders.

Protects Itself Like a Skunk

Bunnies are cute and cuddly just like rolypolies, but both animals have the unusual (some people might say disgusting) habit of eating their own poop. When their preferred foods are scarce, both bunnies and rolypolies eat **scat** (that's a scientific word for poop). Since their digestive systems are not able to break down all of their food the first time through, bunnies and rolypolies can get an extra energy burst by dining on scat. As the material in the scat passes through the rolypoly a second time, more nutrients are absorbed by the rolypoly's digestive system. It's a great way to recycle waste. For a rolypoly, one out of every ten meals is scat. The other nine meals might consist of other appetizing entrées such as fungus, rotten plants, fruit, spider eggs, or any kind of dead animal.

Have you ever heard of people who lack energy because they have iron-poor blood? Iron is an important part of human blood that helps us breathe. When we don't have enough iron in our blood, we feel lazier than sloths. Leafy vegetables like spinach are rich in iron. Perhaps that's the secret of Popeye's strength. Have you ever heard of a rolypoly with iron-poor blood? Of course not! Rolypolies don't need iron for their blood. They instead require copper, the stuff pennies are made of.

Eats Like a Bunny

Without a delicious, nutritious diet of old rotten leaves, rolypolies begin to crave copper. Fresh leaves contain copper, but unfortunately it's locked up in the leaf tissue, where rolypolies can't get to it. Old leaves that have been munched on by microscopic bacteria and fungi are easier for a rolypoly to digest. When rotten leaves are unavailable, scat becomes a must. Since it's been through the gut once, it is chock-full of digestible copper and other nutrients. Carrots, which are available to rolypolies that live in gardens, are even more full of goodies. In an experiment, rolypolies fed carrots but not scat gained weight far faster than rolypolies fed other foods.

Too much of a good thing, however, can spell trouble. A high copper diet can actually poison a rolypoly. Luckily for them, they seem to know when they have all the copper they need.

Like kangaroos, rolypolies keep their babies in pouches until they are big enough to fend for themselves. Unlike kangaroos, they give birth to more than one or two babies at a time. In fact, a young rolypoly mama may bear more than a dozen babies at once, while an older and larger mother may give birth to several hundred. Life must be busy for the rolypoly mother. Imagine having hundreds of tiny wiggling white babies inside your pouch for two to three months at a time, sometimes even twice a year! Luckily your pouch would be protected by the hard upper plates of your first five pairs of legs. Whether you were rolled into a ball or cruising through the strawberries, your young would be safe from enemies and the hot sun.

Rolypoly babies look just like adults except for their small size and light color. As they grow older, their shells become darker each time they shed them, or **molt**. Between the time young rolypolies leave the pouch and develop a harder shell, they live a dangerous life. If the air is too dry, they can die from lack of moisture. If the air is too wet, they may be attacked and killed by fungi. If they get discovered by a spider or a salamander, they are history, because their shells are not yet hard enough to protect them. Those that live to adulthood still don't get very old. Most rolypolies never get any older than a kindergartener.

Since rolypolies live short lives, it's important that they grow up fast. By the time a female rolypoly is a year old, she is ready to become a mother. A female rolypoly is able to produce up to two sets of young from one mating. The young may be both males and females or all the same sex.

Carries Babies Like a Kangaroo

Kid Experiments

Although numerous scientists have studied rolypolies, there are still many mysteries to be solved. Among these mysteries were the questions asked by the kids in the elementary school down the street. Second through sixth grade students at El Portal Elementary School had many questions that could not be answered through reading or observation or discussions with experts. There was only one thing to do in such a case. Experiment! And that's exactly what they did. Here's what happened.

Even though rolypolies #1 and #6 were the fastest, they would have to cruise at that same speed for 22 hours to travel just one mile! That's 44 times longer than it takes most kids to walk the same distance.

How Fast Can a Rolypoly Go?

Caitlin and Rachel set up a racetrack with one-foot rulers and plastic cubes for walls. One at a time, rolypolies raced down to the other end of the ruler. How long do you think it took the rolypolies to travel a distance of one foot? Check out their results.

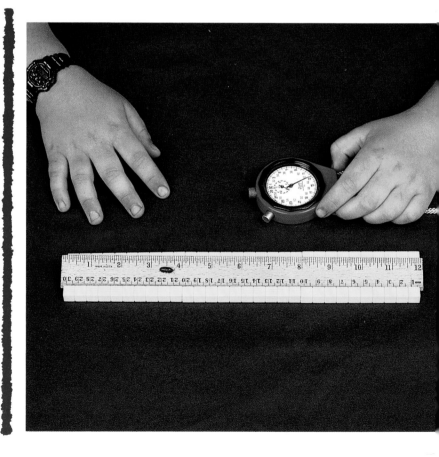

Rolypoly	Time	Rolypoly	Time
#1	15 seconds	#6	15 seconds
#2	30 seconds	#7	25 seconds
#3	25 seconds	#8	25 seconds
#4	45 seconds	#9	30 seconds
#5	35 seconds	#10	30 seconds

How Far Can a Rolypoly Roll?

How far do you think a rolypoly can roll? Tyler, Logan, and Aaron also wondered, so they set up the following experiment. First they leaned a ramp against a bookshelf, and then they created a runway lined with yardsticks at the bottom of the ramp. The ramp was one long half of a four-foot section of foam pipe insulation (which you can get at hardware stores). One at a time, rolled-up rolypolies were released at the top of the ramp and rolled onto the floor. Since the runway walls were made from yardsticks, measurements could be determined quickly by merely looking at the walls. How far do you think they rolled? Check out the results below.

Rolypoly	Distance
#1	22 inches
#2	7½ inches
#3	1½ inches
#4	11 inches
#5	23½ inches
#6	14½ inches

The researchers also noticed that the small ones stayed rolled up longer than the big ones and also went farther. Rolypolies are often observed in the wild rolling into rock crevices where they are safer from enemies. Do rolypolies roll as an escape move? If so, does it make sense that young rolypolies would be better rollers? (See Protects Itself Like a Skunk, page 28.)

Can a Rolypoly Dig?

Josh, a fourth grader, wanted to find out if rolypolies could burrow into soil. He filled a jar halfway with soil, dropped a rolypoly in, and then took the jar outside in the warm sunlight to watch. "He's digging," Josh reported as the rolypoly tunneled its way into the dirt. In less than five minutes, it was totally buried. In the wild, rolypolies commonly burrow to escape harsh conditions such as cold.

Do Rolypolies Like the Dark?

Mandy noticed that her rolypoly liked to go under things, so she placed it on top of a piece of bark and watched it. It went to the edge and crawled to the underside. She turned the bark over, and the rolypoly again went to the underside. Every time she flipped the bark, under went the rolypoly. Later, Mandy noticed her rolypoly crawling into a pencil tip eraser. "They like places dark, so they always live in the dark," she concluded.

Mandy discovered one of rolypolies' basic needs—darkness. Rolypolies lose water through their skin. To conserve water, they avoid hot, dry places and seek out dark, moist crannies.

Can a Rolypoly Go through a Maze?

James and Jacob created a maze using plastic cubes and rods. Included in the maze were several shelters with cubes for walls and larger flat plastic blocks for roofs. Both boys thought the rolypolies might find the shelters and stay inside them, since rolypolies are usually found under rocks or logs. Four rolypolies were chosen as test subjects and then released one at a time at the start of the maze. The first few rolypolies got partway down the maze and then turned around at the first big corner. Eventually, one made it around the corner and proceeded toward the finish. When it got to the last section, it made a right turn instead of the left turn needed to reach the exit, but it soon returned to the junction. At the junction, there was a shelter, but the rolypoly didn't stay in it very long either time. After wandering around for a short time, it found the exit.

James and Jacob noticed that all the rolypolies moved their antennae and stayed close to the walls as they walked through the maze. Most of those that had turned around at the first big corner returned later and found the exit. James and Jacob placed the same rolypolies in the maze four times and timed them each during each test. Here are the results.

Roly-poly	1st time	2nd time	3rd time	4th time
#1	80 secs.	100 secs.	90 secs.	105 secs.
#2	90 secs.	140 secs.	85 secs.	130 secs.
#3	65 secs.	80 secs.	125 secs.	85 secs.

James and Jacob discovered that rolypolies can go through mazes, but they don't seem to learn their way around after repeated trips.

What Do Rolypolies Like Best?

If given a choice of materials to go to, will roly-polies show a preference for one or the other? Kim, Carolyn, and Jessie placed small piles of wet sand, dry sand, wet dirt, dry dirt, and grass on a desk. One at a time, they placed a rolypoly at a fixed starting point an equal distance from the piles. Each girl observed her own rolypoly and recorded results. This is what happened.

Rolypoly	Test 1	Test 2	Test 3	Test 4
#1	wet dirt	wet dirt	wet dirt	wet sand
#2	wet dirt	wet dirt	wet sand	wet dirt
#3	dry dirt	dry dirt	dry dirt	dry dirt

What do you think? Based on this experiment, do individual rolypolies have preferences for different materials?

Can Rolypolies Survive in Water?

What would happen if a rolypoly blundered into a puddle or fell into a stream? Would it live? The bathing abilities of rolypolies fascinated Chelsea and Libby, so they wasted no time in setting up a test. Four clear plastic cups of water were placed on a desk, and one at a time, the girls dropped a rolypoly into the liquid. Here's what happened.

Rolypoly #1:
just lay on its back, alive after 2 minutes
Rolypoly #2:
lay on its side, alive after 4 minutes
Rolypoly #3:
sank to the bottom, dead after 5 minutes
Rolypoly #4:
lay on its back, alive after 2 minutes

Chelsea thought that perhaps rolypolies survived at least a short time in water because they are related to crabs, lobsters, and other crustaceans. Since crabs and lobsters live in the sea, Chelsea hypothesized that survival might be better in salt water. Rolypoly #4 was placed in a cup of salty water and observed by Libby and Chelsea. What do you think happened? It is sad to report that the rolypoly died after 2 minutes. Libby and Chelsea felt bad about the death and decided not to do any more experiments with rolypolies in salt water.

In 1960, Oscar Paris, a zoologist at the University of California, noticed many dead rolypolies after a heavy rainstorm in the Berkeley hills. He wondered if they could have drowned, so he set up an experiment in his laboratory to determine how long rolypolies could stay submerged in water. All of them survived in pond water for at least an hour. Over half of them lived for more than four hours in water. After being removed from the water, the rolypolies flexed their abdomens. Oscar guessed that this was how they squeezed water out of their **pleopods** (their breathing organs). Many of the rolypolies that were submerged for up to seven hours were alive when they got out of the water, but later died from illness. The smaller the rolypoly, the longer it survived. Later, when Oscar surveyed rolypolies after heavy rainstorms, he discovered that as many as one out of ten had drowned.

Why did the smaller ones survive longer? Research often leads to more questions that are left dangling on the tail end of scientific reports. Who will answer them? Will you?

Are there any questions that you are anxious to follow? Can you create any experiments to help you solve muddling mysteries? Sure you can. Just get some rolypolies and do it!

Have you ever wondered what living in a really crowded place would be like? Suppose you lived in a small house with fifty or a hundred people. Philip Ganter, a scientist from North Carolina, wondered what happens to rolypolies when they live in crowds. First of all, he assumed that food would become scarce for rolypolies. To test his idea, he placed different populations of the species *Armadillidium vulgare* in small plastic cages of equal sizes. Each group had a different number of members: six, twelve, and twenty-four rolypolies per cage. Each group was fed the same amount of rabbit chow. What do you think happened?

As you might have guessed, the rolypolies in the most crowded group grew more slowly. What do you think happened when he conducted the same test but changed the amount of food so that all the groups had more food than they could eat?

What Philip discovered didn't surprise him. Even with excess food, growth slowed down in the most crowded cage, though it didn't slow as much as when they had only a limited amount of food available. Why do you think this happened? Philip concluded that rolypolies interfered with each other's eating when there were too many of

Six Is Company, Twenty-Four Is a Crowd

them in a small area. Do you think the same thing would happen if you tried to fit fifty people at your dinner table every night?

After further experiments, some other interesting discoveries were made. When Philip placed two different kinds of rolypolies in the same cage, the growth rate of both kinds was higher than in the previous test. Why do you think this happened? If one kind of rolypoly ate faster or moved more slowly than the other kind while feeding, do you think it might make a difference in how much they interfered with each other? This is what Philip Ganter thought.

In both cases, either in a mixed population of rolypolies or one that was all one species, the growth rate and number of rolypolies went down when conditions were crowded. Philip Ganter considered that this might actually be very helpful to the rolypolies. Philip had learned from Oscar Paris about a large group of rolypolies that had been killed by a fungus. The disease had started by infecting only a few rolypolies, but it quickly spread through the rest of the population because the rolypolies lived in a very crowded neighborhood. In a less crowded situation, the infection might not have been so serious. Philip Ganter thought that perhaps interference during dining helps rolypolies in the end by keeping them from becoming so crowded that they all get sick and die.

No doubt your folks taught you that eating poop could be bad for your health. Too bad for rolypolies that they have never had the same lesson. When rolypolies nibble on starling droppings, they unfortunately swallow the eggs of thorny-headed worms (also known as acanthocephalan worms). These eggs have been deposited by adult worms living in the bodies of the starlings. Like punk rockers, these worms have heads covered with spikes. At the end of the spikes are hooks. That's why they are called thorny-heads. The hooks are used to set anchor in an animal's gut so that thorny-heads can just hang out and steal food that passes by. This kind of thievery is called **parasitism,** and thorny-heads are genuine parasites. Like most parasites, they have deluxe accommodations, but there's not enough room for kids. That is why thorny-heads release their eggs out the bird's back door, encased in bird droppings. When rolypolies munch on starling droppings, guess who gets worms.

Janice Moore, a Colorado scientist, wondered if rolypolies with parasitic worms acted differently from other rolypolies, so she created a series of experiments to find out. In each experiment, she used observation chambers fashioned out of glass pie plates. During each test, she watched two sets of

Rolypolies, Starlings, and Worms

rolypolies in two different chambers. One set had been fed starling droppings. The others were fed food that wasn't infested with worms. All of the rolypolies were marked with paint so they could be kept track of individually.

In one experiment, Janice used a tile to make a small shelter in the center of each chamber. The rolypolies who were not fed droppings went under the tile shelter almost three times more than those fed it. In another experiment, she covered half of the floor of each chamber with dark material and the other half with light material. Rolypolies that had munched on starling droppings spent more time on the light material than the other rolypolies. Finally, in another test, Janice walled off two equal-sized rooms in each chamber with glass dividers. One was kept moist, the other side relatively dry. Scat-fed rolypolies were found in the drier part more than non-scat eaters.

Not only did Janice Moore see that rolypolies parasitized by thorny-heads act differently, but she realized that they seem to become confused. Instead of looking for dark, moist shelters, they end up in open, unprotected places where it is easier to get caught and eaten by starlings, who then end up with worms in their guts. So the cycle repeats itself! To Janice, it seemed that the wrong-way behavior of infested rolypolies is definitely helpful to thorny-headed worms. Do the worms change rolypoly behavior on purpose? That's what Janice wondered.

Like rolypolies, people can also get parasites from poop. Aren't you grateful someone taught you what and what not to eat? Aren't you glad you washed your hands before your last meal? What, you didn't? Uh-oh!

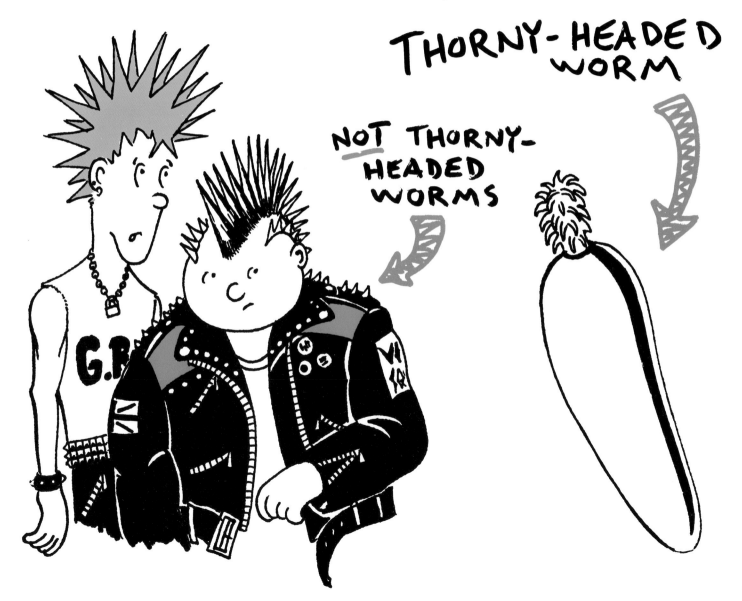

THORNY-HEADED WORM

NOT THORNY-HEADED WORMS

Question Everything!

When you followed your questions, did your journey lead to more questions? Since scientists are continually running into new questions while they ponder old ones, most lead lives of never-ending adventures into the unknown. Even though rolypolies are common little backyard buddies, there is still much to learn about them. Both the kid and adult scientists discussed on the previous pages learned more by pursuing questions. Did you? Questions always lead to answers, even if the answers aren't the ones we seek. Even simple questions seldom lead to simple answers. The kids in my town found lots of new questions as they investigated rolypolies. A few are listed here.

Can people eat rolypolies?

How do you keep rolypolies from eating strawberries?

Can rolypolies survive in outer space?

Why can't the flat ones, sow bugs, roll up?

Can they talk to each other?

Can you train them?

Do they come in other colors?

CAN THEY READ?

Do rolypolies roll from one place to another?

Glossary

abdomen: the second section of a rolypoly's body

antennae: structures on the heads of some animals, including rolypolies, that are used to sense the world around them

antennules: the first pair of antennae on a rolypoly's head

arthropods: a group of animals that includes insects, crustaceans, and other creatures. All arthropods have bodies made up of more than one section and jointed legs.

carapace: a hard covering on the heads of some crustaceans

cephalothorax: the first section of a rolypoly's body, which includes the head

crustaceans: a group of animals with hard outer skeletons. Lobsters, shrimp, and rolypolies are some examples of crustaceans.

exoskeletons: the hard outer skeletons of some animals, including rolypolies

isopods: a group of crustaceans whose legs are all the same size and shape, unlike the legs of other crustaceans

mandibles: the mouthparts of some animals, including rolypolies

molt: to shed the skin or exoskeleton

parasitism: depending on another animal for survival in a way that causes harm to that animal

pleopods: one of an isopod's sets of legs, used for breathing

predators: animals that hunt and kill other animals

scat: an animal's solid waste

segmented: made up of separate parts

zoologist: a scientist who studies animals

Index

About the Author

For the last twenty years, Michael Elsohn Ross has taught visitors to Yosemite National Park about the park's wildlife and geology. Michael, his wife, Lisa (a nurse who served nine seasons as a ranger-naturalist), and their son, Nick, have led other families on wilderness expeditions from the time Nick learned to crawl. Mr. Ross studied Conservation of Natural Resources at the University of California/Berkeley, with a minor in entomology (the study of insects). He spent one summer at Berkeley raising thousands of red-humped caterpillars and parasitic wasps for experiments.

Raised in Huntington, New York, Mr. Ross now makes his home on a bluff above the wild and scenic Merced River, at the entrance to Yosemite. His backyard garden is a haven for rolypolies, crickets, snails, slugs, worms, and a myriad of other intriguing critters.